Welcome to Our Apple Farm

Illustrated by Lance Raichert & Paul E. Nunn

Published by
Pyramid Publishing
P.O. Box 129
Zenda, Wisconsin 53195

A

Welcome to our apple farm !
There are lots of things to learn and do...

Apples come in different kinds and colors.
What colors should these apples be ?

Let's find out where apples come from.
Follow the path to the orchard.

Apples grow on trees.
But how do apples get there ?

An apple tree starts with an apple seed.
The seed needs sunshine and water to grow.

It takes 4-5 years before the tree is big enough to make apples. When it's ready, sweet smelling flowers pop out in the spring.

How many bees can you count ?

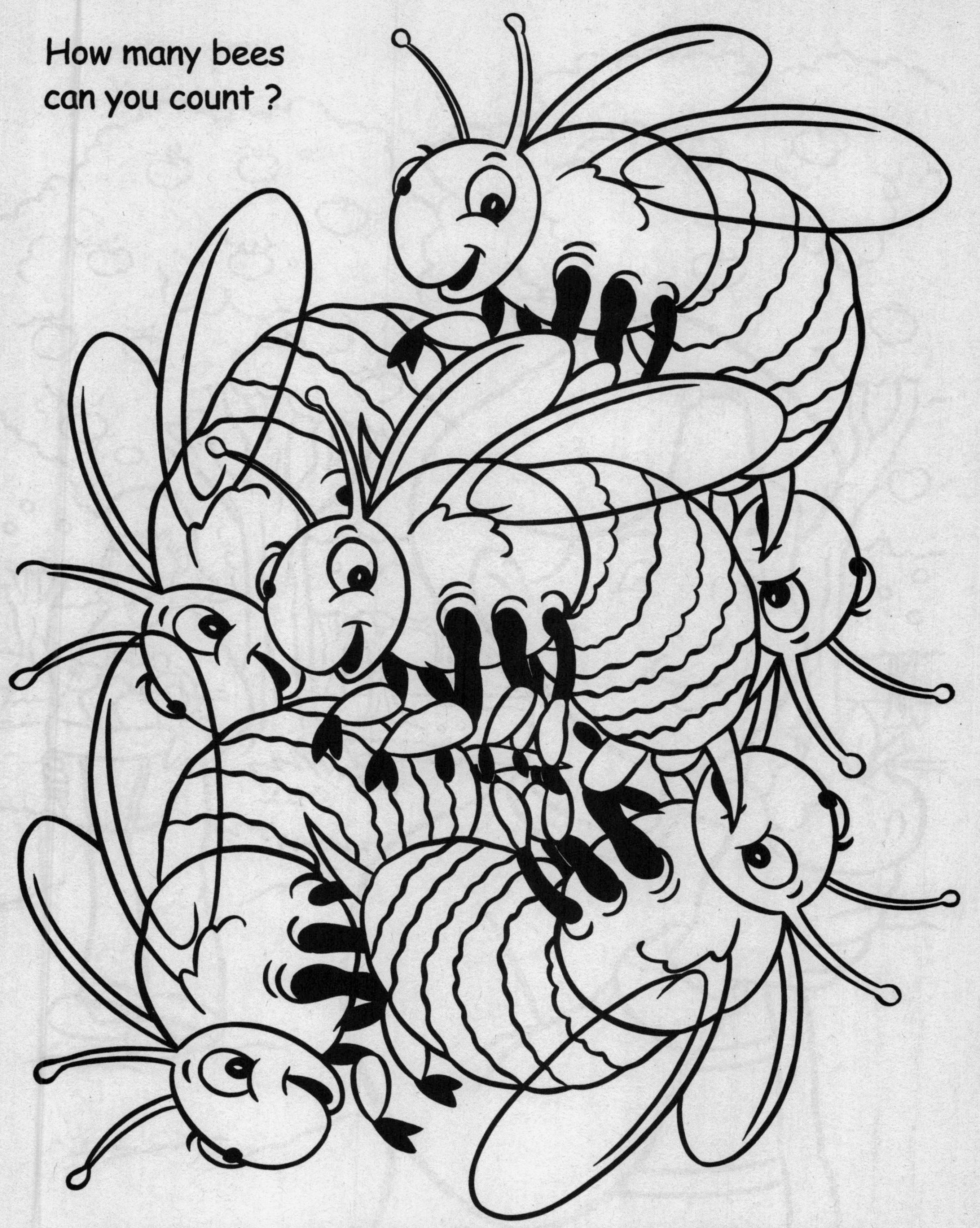

Bees spread pollen from blossom to blossom. The blossoms turn into baby apples, and the apples spend the summer growing.

By fall, most apples are all grown up and ready to pick.
Jack is eating a **Red Delicious** apple.

Some varieties of apples however,
are ready in July

A

What happens to apples after they're picked ?
Let's follow our friends and see.

Something smells good in the apple store!
What do you think it is ?

Mmmm .... It's an apple pie !

APPLE PIE
APPLE PIE
APPLE PIE

Joanna is learning to make apple pie.
Today, they're using **Jonathan** apples.

McIntosh apples make good caramel apples.

Wouldn't you like one of these ?

APPLE PANCAKES
APPLE SAUCE
CIDER DONUTS

Apples are used in lots of things.
Can you think of other ways to use apples ?

Mom and Dad enjoy apple treats.

Apple cider is made from apples....
But how do they do that ?

Let's take a hayride and see !
Follow the path to the cider mill.

Several kinds of apples are mixed together.

After the apples are washed,
they go into the cider press.

The apples are all ground up inside the press, and the juice is separated from the seeds and the pulp. What comes out is - apple cider !

Eat an apple every day. It's a good way to stay healthy.

Jack is tired !

It's time to say goodbye to our friends at the apple farm. But first....

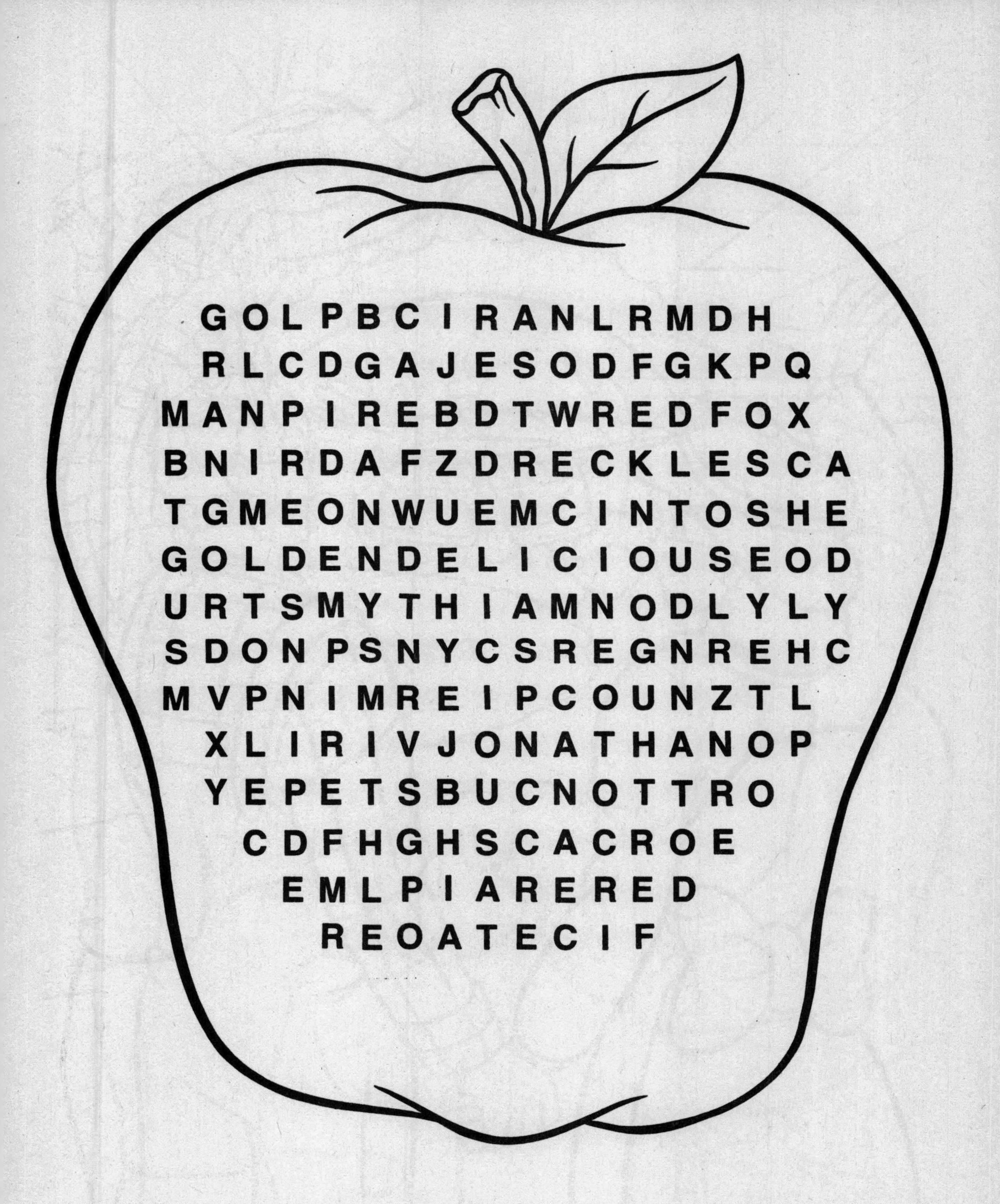

Can you find six kinds of apples hiding in this puzzle ?

EMPIRE, GOLDEN DELICIOUS, GRANNY SMITH, JONATHAN, McINTOSH, RED DELICIOUS.

Thank you for visiting our apple farm.
Come back and see us soon.